WOMEN SCIENTISTS WHO CHANGED THE WORLD

Manuel F. Varela
Ann F. Varela
Michael F. Shaughnessy

STONE TOWER PRESS

Women Scientists Who Changed the World

Stone Tower Press
7 Ellen Rd.
Middletown, RI 02842
stonetowerpress.com

ISBN: 979-8-9992313-0-7

Book design by Amy Cole and Amy Landheer, JPL Design Solutions

Printed in the United States of America

Table of Contents

Introduction

Adolescent girls need more female role models in science. This book attempts to provide young girls, and others, with basic information about some of the women who made enormous contributions to science and society. The book is written for girls who are curious about learning, especially those interested in science, medicine, nursing, famous scientists, and our understanding of certain basic science processes.

We have tried to simplify the book—yet provide a solid understanding of science, medicine, and nursing and the contributions of these females. We encourage girls (and boys, if you are reading this book) to follow their dreams, hopes, and aspirations. We hope to inspire girls to study science in school, read books about it, and learn about the lives of these famous women who have given us so much!

At the end of the book we have included a glossary of terms that you might find helpful with some of the scientific words. We hope that these biographical sketches will inspire and encourage you to follow in the footsteps of these remarkable women.

CHAPTER 1

FLORENCE NIGHTINGALE

1820–1910

SANITATION AND NURSING

Florence Nightingale was a British social reformer and nurse who is known as the founder of modern nursing. She was also important in the field of statistics. Her work training nurses and caring for sick and wounded soldiers during the Crimean War (1853–1856) gained international attention and acclaim for her. Later, in London in 1860, she established the first secular nursing school in the world.

She is well known for her caring for her patients. However, Florence did much more than visit wounded soldiers and sick individuals in hospitals. She was involved in the training of nurses. Back in those days, in England, there were not very many schools for nurses. But Florence decided to train nurses to work with all kinds of diseases, medical problems, and wounds from war and military service.

Nightingale insisted on keeping things clean and neat and wanted sanitary conditions. She believed that germs, like viruses and bacteria, could cause disease and she tried to prevent patients from dying by improving patient hygiene and hospital conditions.

She kept very clean, exact, precise, meticulous records. She looked at numbers, kept good records, and systematically began collecting data and documents. These records helped other scientists discover things about illnesses. She created charts, Venn Diagrams, bar graphs, and other visual means of looking at data. This data collection method helped us learn about the causes and treatment of illnesses and diseases.

She often taught using these visual aids. She constructed what we today call "pie charts" or rose diagrams (because they looked like flowers). These diagrams helped nurses and others understand what was going on with patient responses to treatment and nursing care.

Florence was also involved in moving and turning patients in the bed during the night to prevent what we call "bed sores." Even today, nurses are urged or ordered to transfer or turn patients during their hospital stay to prevent bedsores, which can be very problematic to treat.

Today, we could say that Florence Nightingale is the mother of data-based or evidence-based nursing. If Florence were alive today, she would want to ensure that nurses keep careful records. Today, we all have heard the term "vital signs." A nurse obtains the temperature of a patient, blood pressure, and Pulse Ox (or Pulse oxidation), which refers to blood flow to the extremities of the body, such as toes and feet.

Also, nurses routinely weigh patients—to see if the patient has gained or lost a good deal of weight as this could be a factor in their illness or recovery from sickness.

Nurses (and patients) everywhere owe tremendous gratitude to Florence Nightingale, her work, and her contributions. In modern times, nurses and doctors are trained to follow the teachings of Nightingale, keeping hospitals and clinics clean and free of germs. Physicians and nurses wash their hands, wear masks, and sterilize and clean any instruments, such as a thermometer that takes a patient's temperature. All these things are commonplace and routine today–but in her era, it was not standard procedure. Florence Nightingale did much to make nursing a profession. In so doing, she elevated the status of women who were nurses–one of the very few opportunities that women had open to them in her time period. Today, there are not such limitations on what girls and women can do for work and professionally. Today, if you dream it, you can do it–and in part, that is due to the wonderful and inspirational work of Florence Nightingale.

CHAPTER 2

ANGELINA FANNY HESSE

1850–1934

AGAR AND MICROBES

For those wanting to become scientists, you need to know about items routinely used in a laboratory, such as the Petri dish. Many chemical and scientific experiments are conducted in these Petri dishes, invented by Robert Julius Petri. Another thing commonly found and used in a laboratory is "agar," a substance made of seaweed. Thanks to the ingenious insight by the American scientist Angelina Hesse, scientists worldwide use agar to learn about germs we call microbes or microorganisms. Scientists prepare Petri plates with agar and grow microbes in incubators, making learning about these tiny living creatures easier.

Agar helps scientists to study microbes by growing them on its surface. As these microbial creatures grow, they wiggle, move, eat, and produce new substances. Agar makes these microbial activities possible so we can study them in a laboratory and the classroom.

Before agar, scientists used gelatin, the same substance as is used in Jell-O, to learn about microbes. Gelatin was good for growing microbes at room temperature. When solid, the gelatin allowed the microbe to form colonies on its surface in Petri dishes.

Unfortunately, there is a problem in the use of gelatin in a laboratory when studying microbes—it melts at the "blood heat" temperatures! Disease-causing microbes grow best at high temperatures, like our body heat, 98.6 degrees Fahrenheit (98.6°F) or 37 degrees Celsius (37°C, Celsius is also called centigrade). Gelatin melts at this temperature, preventing microbes from forming colonies and, thus, preventing scientists from studying the blood-heat-loving microbes.

At first, Angelina Fanny Hesse used agar as a food thickener in her kitchen. Agar was used by many for adjusting the consistency of foods, such as soups or ice cream. In the lab, Hesse suggested using agar instead of gelatin. The agar stayed solid at the blood heat temperature. When Hesse used agar, it solved the melting problem of using gelatin. She and others used agar to cultivate bacteria and other microbes.

Later, agar was used to study tuberculosis. Tuberculosis is a dreaded and fearful coughing disease of the lungs. The microbe grows best at the body temperatures of the lung.

Another scientist, German physician and microbiologist Robert Koch (1843–1910), studied the illness and won the Nobel Prize for finding the microbe responsible for tuberculosis. He used Hesse's agar process to find the tuberculosis microbe (though he did not credit her in his publications).

The Nobel Prize is one of the highest honors a scientist can receive. It is awarded every year in Sweden in a small place called Gamla Stan. As we will read, many female scientists in this book have been awarded the Nobel Prize for their scientific contributions.

Angelina Hesse was also known for her scientific pictures, drawings, and illustrations. She had a gift for drawing realistic representations of microbes using watercolors. Many of Hesse's drawings were used in scientific journals. These drawings by Hesse can still be found in libraries today.

Most scientists use agar in their microbial studies to learn about cells and tissues. Thanks to Hesse's use of agar, we know so much more about microbes than would otherwise be possible. We also know much more about our cells and how they work. Scientists also know more about cancer and viruses and how to treat other diseases because we can grow cells on agar.

CHAPTER 3

ANNA WESSELS WILLIAMS

1863–1954

PIONEER OF VACCINES

Female scientists and physicians have done much work in the treatment of many diseases. Dr. Anna Williams was one such woman. She helped treat diphtheria and helped in the diagnosis of rabies (many of you have heard that dogs have to get a "rabies shot" to prevent them from getting the disease). Anna also studied the factors and aspects that led to scarlet fever, influenza (the flu), and trachoma, a bacterial infection involving the eye that sometimes leads to blindness. All of the above medical conditions are very serious. Her work in developing vaccines, treatments, and diagnostic tests for these diseases was extraordinary.

Diphtheria is an illness of the respiratory system or breathing system. A germ bacterium causes this disease. In her day, patients with diphtheria had a sore throat and a slight fever and later died of shortness of breath—a common thing in patients with heart troubles.

Unfortunately, when given food and water, the microbe grew much too slowly to study in the lab. Seeing it unfold on Hesse's agar in Petri dishes took weeks.

In 1894, Williams made a giant discovery. She found a "strain" or an example of the diphtheria germ. The strain had special properties that were useful to her.

The new strain discovered by Williams was a fast-acting germ, making it helpful to grow better. The faster-growing microbe permitted Williams to study the bacteria better. The germ discovery by Williams created a path to getting a vaccine or treatment for preventing diphtheria.

Today, thanks to Williams, the diphtheria illness is seen much less often where the vaccine is used. Williams also was involved with developing a way to diagnose "rabies," which for many years was unavailable. Williams discovered a way to grow the rabies germ, a virus microbe, in the lab. She used a weakened version of the grown microbe to make a new rabies vaccine.

She also studied a sickness called "trachoma," which can lead to a person becoming blind, especially in children. Often flies are blamed for transmitting this tropical disease.

Williams found a new way to grow the trachoma microbe, a bacterium. Williams used the grown trachoma bacteria to invent a lab test. The trachoma test was used to screen children in school to protect them from sickness and blindness. Nowadays, children are tested or screened for vision problems.

Williams had been interested in the microscope since she was young. At 12 years of age, she peered into a microscope and became forever fascinated, inspiring her to study science. After the near-death experience of her close sister, Millie, Williams dedicated her life to becoming a medical scientist using the microscope.

Williams developed a new scientific method of diagnosing rabies. First, she smeared brain slices from dogs infected with the *Rabies virus* onto microscope slides. Then she stained the rabies-infected brain slices with chemicals to see the infection better. The new method was better than all other lab tests, and it was used for over 30 years by medical scientists to diagnose rabies.

Anna Wessel Williams worked in the New York City Department of Health for 39 years. She devoted most of her life to helping sick individuals. Because of her dedication to science, many people have been free of several otherwise terrifying diseases.

CHAPTER 4

MARIE SKLODOWSKA CURIE

1867–1934

RADIOACTIVITY AND X-RAYS

Although many students of science and medicine recognize the name Dr. Marie Curie—her full name was Marie Salomea Skłodowska Curie. And she deserves a lot of recognition and acknowledgment. She is important in science and medicine because she is the first person to receive two Nobel Prizes. Curie was the only woman to receive two Nobel Prizes. She is buried at the Pantheon in Paris, France, another great honor.

Curie was the first female professor, lecturer, and Head of a Laboratory at the Sorbonne in Paris, France. The Sorbonne (Sorbonne Université) is a famous university in Paris, France. Many people go there to study and learn from some of the world's leading scholars. She was also the first woman to teach at the Sorbonne.

Marie Curie was the first woman scientist Nobel Prize laureate. The Nobel Prize is a tremendous honor given to people who make outstanding contributions in their respective field of research. Her first Nobel Prize was in Physics (1903) for her discovery of radiation. Curie's discovery of new elements, polonium, and radium, was worthy of a second Nobel Prize, this time in Chemistry (1911). Curie also discovered the atomic element thorium. The energy radiates from the insides of some atoms, like uranium. We call this energy radiation because it is released from such atomic particles. Thorium is one of those elements.

As a young woman, she was fascinated with X-rays. During World War I, Curie designed mobile X-ray units—trucks with portable X-ray machines. They were known as "radiological cars" and Curie and her teenage daughter Irène drove the mobile units to take X-rays of war wounded in France to help treat them. Today, Curie's work with X-rays is the basis of radiology and medicine that is used in almost every hospital worldwide.

Her interest in X-rays led to her discovery of different energies radiating from atoms! All matter in the universe is made up of tiny atoms. At the time, atomic bits were thought to be unbreakable. She discovered that some of these atoms emitted radiation, a new form of energy. Curie invented the word "radioactivity" to describe that new energy.

Curie came from a family of scientists. The Curie family enjoyed learning at home, reading books, and absorbing literature, music, and art. Her father, Vladislav Curie, was a professor of physics and math. Her mother, Bronislava, was a teacher and director of a school for girls. Curie and her family continued to contribute to science for many years—in many ways.

Marie Curie also influenced others in her family—especially her daughters Irène and Eve. Like her mother, Irène also studied physics and chemistry. Physicists study matter and energy. The universe's matter makes up all substances in the cosmos. Chemists study how matter interacts to form new substances.

Irène and her husband, Frederic Joliot, worked together and discovered how to make the radioactive materials Marie had found. For this later discovery, Irène and Frederick would get the Nobel Prize.

Curie's second daughter, Eve, would affectionately write a biography of her world-famous mother, Marie. Eve and her husband, Henry Labouisse, Jr., would receive the Nobel Peace Prize for their work at the United Nations Children Fund (UNICEF). The UNICEF headquarters is in New York City, and the organization tries to help people worldwide, especially children with diseases and illnesses.

Marie Curie worked tirelessly as a scientist and a humanitarian. She was the first female scientist to become an international celebrity. She truly was a person committed to science, and she never allowed her gender to be a hindrance. She truly was an exceptional scientist and her work changed the world for the better.

CHAPTER 5

ALICE BALL

1892–1916

RESEARCHING LEPROSY, A GROTESQUE AFFLICTION

Although a person named Gerhard Hansen was the person who discovered the germ that causes leprosy, it was African American chemist Alice Augusta Ball who made many discoveries that led to its effective treatment.

Alice was born in Seattle, Washington but when she was still a child her family moved to Honolulu, Hawaii. After a year the family moved back to Seattle and Alice eventually studied chemistry and pharmacy at the University of Washington. She then returned to Hawaii to pursue a master's degree at the University of Hawaii. She was the first woman and first African American to receive a graduate degree there, and she was also the first female chemistry professor at the university.

Leprosy is a very contagious disease, which means it is easy for those afflicted to transmit it. Also called Hansen's disease, leprosy is a terrible disease caused by a microbe. Patients with the illness get skin rashes which can be severe, causing horrific disfigurements. Historically, patients were forced to live in separate "leper colonies." In Medieval times, those afflicted with the illness had to carry clappers (bells) and sound them to warn others that they were nearby. The only effective treatment for Hansen's disease was found in a tree known as a chaulmoorga tree.

Alice Ball worked with a leper colony in Hawaii for some time. She became aware of the potential cure for Hansen's, the seed oil of the chaulmoogra tree. The chaulmoogra tree—is found in tropical forests along the Western Ghats mountain range in India and Southeast Asia and contains a seed oil that lessens the sores in Hansen's disease patients.

Scientists failed to isolate the curative oil from the tree for decades. The problem was that the chaulmoogra oil was gummy. That is, the oil cure was too viscous or dense to be used.

Viscosity refers to the thickness of fluids, like the texture of honey. Alice Ball used her skills and chemistry training to invent a technique for removing oily chemicals from the tree. She also discovered a way to make the medicine less gooey so that it could be used to treat Hansen's disease patients. It was arduous work, and only Alice Ball could complete the curative chaulmoogra oil medicine production.

Sadly, before the cure was applied to human patients to cure them of Hansen's disease, Alice died at the young age of 24, possibly from a laboratory accident. The cause of her death is not clear. Some say Ball had inhaled chlorine gas, which poisoned her by accident. Others say she succumbed to the dread disease at the time called "consumption," a lung disease caused by inhaling a bacterial microbe. Her original death certificate was altered to read that she died of tuberculosis. No one knows for sure what really killed her.

Ball gave her life to science and the treatment of leprosy. We can never know what other important discoveries Alice Ball might have made had she lived longer than 24 years.

CHAPTER 6

RUTH ELLA MOORE

1903–1994

BLOOD, TOOTH DECAY, AND TUBERCULOSIS

Ruth Moore was the first African American to earn a coveted doctorate in the natural sciences. She went to Ohio State University to earn a Doctor of Philosophy degree (Ph.D.) in 1940.

For her doctoral project, Ruth was concerned about tuberculosis and the field of immunology. She studied how the bacteria cause tuberculosis, a lung ailment in which patients cough excessively, often dying from their illness. In her era, the sickness was so terrible that those who fell ill to it were described as having "The consumption"—that is, having a disease that consumed them. Moore devised a technique for measuring the microbe in fluid specimens from tuberculosis patients.

Immunology is the study of what makes people resistant (immune) to disease. The bacteria that cause tuberculosis are dangerous because they avoid the immune system, allowing the patient to be consumed. The bacteria escape from white blood cells that control invading microbes. Instead, the tuberculosis-causing microbes provoke or bring on pneumonia, an inflammation of the patient's lungs.

Ruth Moore taught at a prestigious college known as Howard University and published a very important article on tooth decay and cavities, or dental caries. She predicted that bacteria in the mouth could ferment sugars into tooth-decaying acids. As an experiment, Moore injected mouth bacteria-detecting substances, called antigens, into the skin of human volunteers and detected more swelling. Today, dentists worldwide owe her a debt of gratitude for her work on tooth decay.

Ruth was one of the earliest known scientists to study the blood types of African Americans. She discovered that the numbers of blood types did not differ much from non-African American populations. Moore confirmed that the common blood type was type O for the two populations of people.

Doctors and nurses need to know the blood types of their patients so they can be treated effectively. We know that all humans share similar blood types thanks to Moore's pioneering blood studies. People who have type A blood have on the surfaces of their red blood cells a protein called the A antigen. Likewise, in type B blood, people have the B antigen on their red blood cell surfaces. Some people have A *and* B antigens on their red blood cell surfaces. Type O people have neither A nor B antigens in their blood.

Type A people can receive blood transfusions from type A or O. These recipients can make blood-destroying antibodies to type B or AB blood. The immune system is too powerful and dangerous to patients receiving incorrect blood types. They cannot receive incompatible blood.

Type B blood recipients can have type B or O. These individuals produce antibodies to type A and AB blood, which are incompatible with type B.

Type AB individuals can receive blood from type A, B, AB, or O. They do not produce antibodies to these blood types. They are considered "universal recipients."

Type O people are considered "universal donors." Blood type O does not elicit antibodies from recipients.

Another blood factor considered by Moore is called Rh Factor. All people either have or lack these Rh factors. Those who have it are called Rh+, and those without are Rh-. Thus, blood transfusions can be more complicated if one is, for example, A+ or A-. These and all other individuals have Rh factors to consider when receiving or donating blood.

In a completely different field of study, Ruth also learned about bacteria that live in the guts of cockroaches. She discovered that many of the gut microbes were unaffected by medicines. She followed up on this breakthrough by studying whether medications stopped bacteria from growing or killed them outright. These studies were important because these insects can transmit disease-causing microbes by crawling around houses and food stores, and the diseases may resist treatment.

Apart from her pioneering scientific achievements Ruth was also an accomplished seamstress and fashion designer. Today, she remains a role model for anyone studying or working in the field of microbiology. She was determined, resilient, and as a teacher, she trained thousands of students in science and medicine. Her legacy lives on in laboratories and the lives of scientists around the world.

CHAPTER 7

ILONA BANGA

1906–1998

VITAMIN C

Today, almost everyone knows about vitamin C and its importance to health and the human body but that was not always true. Many of us have heard about vitamin C and its name—ascorbic acid. But who was one of the first scientists to study and investigate it? It was Ilona Banga. Banga was a colleague of Hungarian biochemist Albert Szent-Györgyi and contributed to his work that earned him the Nobel Prize in Physiology in 1937. During World War II, she worked to keep her lab safe from its destruction.

Banga was born in southeastern Hungary in 1906. As a student, she wanted to study to become a doctor but her mother thought that was not a good choice for a woman so she chose chemistry instead. She received a Master of Science degree in chemistry from the University of Debrecen in 1929 and soon after began her career as a research assistant to Albert Szent-Györgyi at the University of

Szeged's Institute for Medicinal Chemistry. It was there that she did her important work on vitamin C.

Our cells need vitamin C to make the energy for living. Vitamin C is important because the human body does not make it naturally. So we either must eat a lot of oranges or other fruits and vegetables with vitamin C or take a pill so that the human body gets enough of this substance. Her work made an enormous contribution in understanding how vital vitamin C is for the human body.

Banga studied paprika and found that this pepper contained a large amount of vitamin C. Much experimental work went into isolating vitamin C from paprika. The effort was complicated and demanding but Banga succeeded in preparing pure samples for further study. This was a significant breakthrough because no one had ever seen vitamin C in a pure form. Banga was the first to make this discovery.

Many years ago, there was a disease called scurvy. Without vitamin C, humans get this sickness, which can be serious. In previous centuries British sailors (and others) on long voyages would often become quite ill because of their diets that often lacked vitamin C. After a while, it was found that they did not become sick by eating limes because they contained some vitamin C. Thus, the sailors acquired the nickname "Limeys."

The substance has been used to make humans have healthier lives. Thanks to the work of Banga, the scourge of vitamin C deficiency has diminished greatly in places where foods are fortified with the vital agent.

Banga also studied the human body's muscles and how we, as humans, age. Today, we refer to this field of study as "gerontology." Humans need more vitamins, nutrients, proteins, and minerals as they age.

Banga studied how muscles use energy to contract. She made pure muscle proteins, placed them in a test tube, added energy power, and boiled muscle juice. Banga watched in amazement when the proteins contracted like muscles in the test tubes!

Banga was also interested in arteries that carry blood in the body as it ages. She learned how blood arteries stretch and shrink back by studying a protein that permits them to be elastic. During aging, old blood arteries become stiff, making it problematic for the heart to pump blood, increasing blood pressure. High blood pressure can be unhealthy and that is why one of the first procedures nurses and doctors perform when seeing a patient is taking a patient's blood pressure.

Sometimes, the heart arteries can become blocked by plaques of fat and cholesterol created by an unhealthy diet or by a rare genetic defect in which patients make too much cholesterol. These situations can lead to heart attacks when the blocked arteries cannot bring oxygen to the heart, resulting in the heart tissue dying.

To learn about stiffening aging arteries, Banga studied their proteins. Some proteins in the body can act as enzymes, making chemical reactions occur very fast. These enzymes are named with the suffix "–ase," and Banga's enzyme was called elastase.

Elastase is a protein that was studied closely by Banga. It was the earliest enzyme to be isolated by a scientist. She found that the elastase enzyme could degrade fibers in the body called elastin, which lined the veins that carry blood. She studied elastase in people with a disease called arteriosclerosis. The condition is characterized by fat (lipid) deposits in arteries and veins, limiting the blood flow through the body.

Banga found that patients with clogged arteries had lower elastase amounts than did healthy people. These arteriosclerosis patients had a greater risk of stroke, heart attacks, and death. Thanks to the pioneering works of Banga, we can avoid conditions that increase these risks and live longer and healthier lives.

CHAPTER 8

MILDRED COHN

1913–2009

RADIOACTIVE ISOTOPES

Dr. Mildred Cohn was a famous scientist who used "nuclear magnetic resonance," or NMR to measure the energy signatures of individual molecules. NMR and magnetic resonance imaging equipment enables doctors to look inside the human body. It is often used daily around the world to make better diagnoses and treatment plans for patients.

Mildred studied human and animal cells to better understand how energy was generated in them. She was interested in special biological power called ATP for living cells to survive and work properly. The term ATP is an acronym for adenosine triphosphate. It is a biochemical made by the body after food is eaten. It is an energy-carrying molecule found in all living things. When the ATP energy is used by cells and tissues to conduct their living processes, the ATP is broken down into smaller pieces. Cohn was able to use technology to advance medical understanding and treatments.

Cohn used the NMR machine to detect ATP undergoing chemical change in the living body. It was the first time somebody had seen ATP and its NMR signature.

The discovery was exciting for Cohn and scientists who read her journal articles about it. Dr. Cohn became world-famous for this exciting scientific discovery.

Cohn was also interested in oxygen, a molecule needed for all human life. She made it radioactive to trace its atoms as they moved from one molecule to another. Cohn was one of the first scientists to use radioactive isotopes to study life, and she was the first to use radiation to examine oxygen.

Cohn received her Ph.D. at Columbia University in New York City in 1938. It was an amazing success story in overcoming barriers to education.

Mildred Cohn had a very difficult life and career because she was Jewish. During the period in which she lived, there was significant "Anti-Semitism," which refers to people who disliked Jewish individuals. During the same era, Cohn also had to deal with those who did not like women entering the scientific world. She conquered both obstacles and earned her college degrees, including the highly sought-after doctorate.

Cohn also studied magnesium. Many adults take magnesium, which is believed to be an essential mineral ion and is needed for life. Cohn discovered that magnesium binds to the cell's energy molecule ATP, helping to make it stable for its chemistry to proceed and allowing cells to live.

Because Cohn studied magnesium, scientists were able to easily measure its amounts in the brain. We now know that magnesium plays critical roles in the brain's development and function, like memory and learning.

Later in life, Cohn studied muscle-related illnesses. Many of these "muscle diseases," such as muscular dystrophy, involve skeleton muscles. This severe condition involves losing the ability to contract muscles, leaving them to shrink, a condition known as dystrophy. Other types of muscle disease involve faulty heart muscle. Cohn used the NMR machine help her understand these ailments. She experimented on muscle chemistry in living cells, a field of study called biochemistry.

Cohn served as the first woman editor of the *Journal of Biological Chemistry* and served on the editorial board of the journal for ten years. This journal is among the most prestigious in all biological sciences.

Before her death, she was inducted into New York's National Woman's Hall of Fame. She was presented with the National Medal of Science by President Ronald Reagan in 1983. Thanks to Dr. Cohn, scientists worldwide understand how living cells work in contracting muscles, how brains function in learning and memory, and how oxygen is involved in the body's chemistry. She is a true role model for women.

CHAPTER 9

JANE C. WRIGHT

1919–2033

"THE MOTHER OF CHEMOTHERAPY"

For years, cancer was a dreaded disease affecting many different body parts. We do not completely understand why cancer strikes people. We know that certain factors, such as smoking, seem related or, as we say, "correlated" to cancer, but there is still much that is unknown.

In past years, until the 1940s and 1950s, cancer was treated by radiation only. If you have read the chapter on Marie Curie, you know a bit about radiation and the work of Marie Curie. For many years, the only routine "treatment," if you will, for cancer was radiation—until the pioneering medical work of Dr. Jane Cooke Wright.

She began to believe there could be other approaches to treatment and cancer cures. She favored chemotherapy—using chemicals to treat cancer. Wright believed that chemicals might be the "Cinderella" of cancer treatment and that their potential was not fully understood.

Before Wright's work was known and became standard practice for the treatment of cancer,, using chemicals was considered useless or dangerous because their successes were highly unpredictable and often prone to failure. Indeed, early chemicals included arsenic, clay, and violet leaves. However, many scientists scoffed at or made fun of these chemicals for treating diseases.

Wright turned this attitude around by using the scientific method of proposing ideas, designing good experiments, and testing the ideas extensively in the lab. She collected lots of data on disease numbers in patients who were likely to die of cancer.

Wright focused on arsenic, which many knew was poisonous. However, Wright brilliantly deduced that an altered form of arsenic would be non-poisonous for humans, but kill cancer cells. She tested the arsenic derivatives on cancer cells in the laboratory, and they worked to extinguish a form of blood cancer called leukemia.

Next, Wright treated very ill patients with the new chemicals and found that such patients treated did better and recovered from their cancers. Soon after her major discovery of chemotherapy, she and many other scientists found newer and better chemicals to treat cancer. The scientists found it immensely important to keep track of the dose-response relationship between the chemotherapy treatments and normal cells versus cancer cells. They wanted to know the relationship between the dose of the drug and the percentage of cells killed by it.

Wright also discovered that she could combine more than one chemical at a time with cancer patients to make the combination treatment work better than each chemical alone. It was another brilliant breakthrough in cancer science and therapy. Thanks to Wright, the marvel of chemotherapy prevailed like Cinderella, and we still use certain chemicals to treat cancer today. We call this chemotherapy.

CHAPTER 10

ROSALIND FRANKLIN

1920–1958

DNA AND VIRUSES

Dr. Rosalind Franklin was a gifted chemist who played a significant role in solving the structure of DNA, the secret blueprint of life. DNA denotes deoxyribonucleic acid, a molecule that codes for all living beings, from germs to humans. Since Franklin's studies led to understanding its structure, DNA has been important for many aspects of our lives. DNA is used to diagnose diseases, solve crimes, correct genetic defects, and study heredity.

Since her childhood in London, she had always been considered exceptionally brilliant. While earning a doctoral degree at the University of Cambridge, Franklin learned a scientific technique called X-ray crystallography to study coal so that it could be used for energy. The method later helped her to solve the mysterious puzzle of DNA's structure. Franklin made DNA crystals and "zapped" them with X-rays, which bounced off the DNA at specific angles. Franklin was already very close to understanding DNA on her own.

The DNA structure was solved from these X-ray diffraction angles when her data became known to other scientists before she had time to solve it independently. Franklin was very clever in that she was the first scientist in the world to prepare two important types of DNA fibers, the A-form and the B-form. When she exposed these DNA structures with X-rays, they bounced off (diffracted) and made spots on photographic films.

She would never know that key aspects of her scientific data were seen by others and used to beat her to the DNA structure puzzle. Before she could do it, these other persons figured out the DNA structure using her data. Moreover, they spoke ill of her to many who would listen, and she would never know about it.

Franklin also studied RNA, ribonucleic acid, used as a transcript to make proteins inside cells. Using her expertise in X-ray diffraction, Franklin studied a famous virus called the tobacco mosaic virus. The virus made mosaic-shaped lesions on the leaves of tobacco plants, ruining them. The last work she ever did was on poliovirus structure.

In the laboratory, Franklin was considered by her close friends and family as a warm and friendly person. Her close associates who worked with her daily confirmed Franklin's gracious character.

Unfortunately, before she could finish her studies on poliovirus, Franklin succumbed to cancer and died in London on April 16, 1958, at age 37. She was not given credit for her DNA data until years after her death. If Franklin had lived longer, she might have been the recipient of a Nobel Prize for her ground-breaking studies of DNA structure. Historians of science speculate that Dr. Franklin would have received additional Nobel Prizes, one for her research on the tobacco mosaic virus and another for the poliovirus structure. We owe abundant gratitude to this scientist for her work!

CHAPTER 11

ELIZABETH BUGIE

1920–2001

STREPTOMYCIN

For many years, penicillin was a "wonder drug," like aspirin, that worked on several conditions. However, some people are allergic to penicillin for whatever reason. There seemed to be side effects to that medication. People allergic to penicillin had to find other medicines to treat their ailments. Thanks to the scientific work of Elizabeth Bugie this new discovery happened!

After graduating from high school, Bugie attended the New Jersey College for Women and concentrated on studying microbes. After college, Bugie went to graduate school at Rutgers University. She worked with a famous scientist named Selman Waksman, participated in a historically significant scientific discovery, and received a master's degree. She had found a new medicine that killed bacteria!

Elizabeth always had been fascinated with science. She was aware of this and sought to develop other variations or variants of penicillin to help treat and cure certain medical conditions. She worked in a well-known laboratory called the Waksman Laboratory. In those days, much research was conducted in the laboratories of prominent male scientists, such as Selman Waksman.

Often women did not get the recognition they deserved, and the head of the laboratory took credit for women's discoveries or their work. Women thus did not get the acknowledgment that they deserved. In modern times, Bugie is widely considered an official explorer who isolated streptomycin, a very important antibacterial drug, from microbes.

What exactly is streptomycin? It is an important medicine for treating diseases caused by bacterial germs. Bugie did most of the experiments in the laboratory to find the chemical from a microbe called fungi grown in large vats (tanks). She tested how powerful the streptomycin was against bacteria in Petri dishes. We know now that streptomycin stops bacteria from making protein, which keeps them from growing and from causing diseases in human beings.

Bugie's laboratory isolation of streptomycin was life-changing for many people. The new medicine could treat patients suffering from what was then called consumption, a debilitating disease that had been around for centuries. Bugie's streptomycin breakthrough could provide needed relief and prevent many patients from dying.

Elizabeth was also a key scientist in discovering other medicines that could be used to treat bacterial infections. She dedicated her life to studying how to kill germs so that patients with illnesses caused by microbial infection could survive their ailments. She told her children that she enjoyed science and found fulfillment in helping to make great scientific discoveries.

Virtually all matter we know of consists of chemicals. We need to understand chemistry to understand how life works or to make new products we need, like food and drink. Chemistry is important in industry, commerce, and technology. The study of chemistry is an critical area when one wants to become a scientist.

If you study chemistry in high school and at the college level, you can learn about the structure of certain chemicals. You also will learn how various chemicals combine to form new materials—hopefully to the benefit of science and medicine.

CHAPTER 12

MARIE DALY

1921–2003

NICOTINE AND CHOLESTEROL

Dr. Marie Maynard Daly was the first African-American to receive a doctorate from Columbia University in New York. Marie Daly was a biochemist, who, like many other scientists, was interested in many things. Scientists are curious people, which helps them develop a passion for learning. Passionate, curious scientists can contribute to society, making the world better. When Daly was a young girl, she read an influential book, "Microbe Hunters," and became tremendously fascinated with tiny microscopic organisms, such as bacteria.

In her research, Daly began to look at certain factors or variables that seemed to be linked to various diseases. Today, we know about the harmful effects of nicotine, which is in tobacco and crosses the blood-brain barrier that protects the brain. She participated in pioneering studies concerning the effects of cigarette smoke. The results of her studies were clear. Nicotine and cigarette smoke damage the lungs and also affect other parts of the body.

What *is* nicotine, and from where does it come? Nicotine is a plant chemical, and it is found in tobacco plants. When people smoke tobacco or consume tobacco products, the nicotine helps them become addicted, meaning that the urge to keep smoking is very strong. It can be hard to stop smoking after one becomes addicted. It is best never to start smoking and avoid the unpleasant effects of addiction and painful lung cancer.

Soon, scientists would discover that the same body damage happened to people. In modern times, we know that cigarette smoking is linked to severe lung cancer in humans and is a top killer disease among people.

Daly also was interested in how certain chemicals contribute to the digestion of food. She was particularly interested in an organ in the human body known as the pancreas. Daly's doctoral dissertation focused on chemicals released when enzymes from the pancreas breakdown corn starch, a storage form of sugar in corn.

Speaking of food, Daly discovered a link between sugar and a substance known as cholesterol. She was one of the earliest scientists to find that cholesterol in the diet could clog blood vessels. Artery blockage leads to abnormally high blood pressure. This condition is called hypertension and is linked to heart attacks.

Marie Daly learned that sugar was sometimes not adequately cleared from the blood, leading to another condition called diabetes. Daly looked at the relationship between diabetes and high blood pressure. She found that sugar was somehow connected to diabetes and high blood pressure.

We probably have all seen doctors, nurses, and other medical professionals using blood pressure machines. What exactly are they doing, and what do they find out? When blood pressure is measured in people, the forces on the walls of the blood vessels are found.

The blood pressure machines detect two numbers. One is the blood pressure on the artery walls when the heart beats, called "systolic." The other number is "diastolic," which measures the lowest pressure between the heartbeats. Healthy blood pressure is about 120 over 80 pressure units.

What exactly *is* diabetes? And are there different types? Diabetes is a condition with too much sugar in one's blood vessels, making the person feel quite sick. A hormone in the body called insulin normally keeps blood sugar levels in a healthy state. But sometimes, the body's insulin cannot do its job, and the blood sugar increases, causing diabetes.

Two categories of diabetes are type 1 and type 2. The type 1 kind sometimes starts in children. It is due to the body attacking its insulin-making cells of the pancreas organ where the insulin is found. Diabetic patients often require an injection of insulin daily. Type 2 diabetes often occurs in grown-ups and may be connected with diet, smoking, or high blood pressure but often with obesity. Treatment also involves, among other things, insulin shots.

How can we prevent diabetes? There are well-known prevention strategies for type 2 diabetes. They include keeping a healthy weight, losing extra weight, keeping your body active by moving or exercising regularly, eating healthy foods, staying clear of foods with high sugar or fats, and avoiding smoking.

Marie Daly's work in science as well as working to advance minority representation in education and science was instrumental in encouraging others to follow in her footsteps. She was a remarkable woman and a remarkable scientist!

CHAPTER 13

ESTHER ZIMMER LEDERBERG

1922–2006

GENETICS

Dr. Esther Zimmer Lederberg was a pioneering scientist who made great discoveries about genetics. She was interested in microbes called bacteriophages. These germs are viruses that infect bacteria and eat them. Thus, bacteriophages are bacterial eaters. As these phages eat, they explode bacteria into pieces, killing them, a process called "lysis." Viral microbes have a genome code covered by a protein shell called a capsid. Because viruses lack the cellular ability to make more viruses, they must steal the mechanisms that do make viruses from living cells.

Lederberg graduated high school at 16 and earned her degree from Hunter College in New York City at 20. She moved to California, where she attended Stanford University and earned her master's degree. Esther then attended graduate school at the University of Wisconsin, where she made a ground-breaking discovery and received a Ph.D.

She overcame many barriers to getting an education. She suffered from poverty, college advisors who discouraged her, no funding for graduate school, and sometimes no food. She and her roommate sometimes resorted to eating frog legs left over from teaching lab experiments!

In graduate school, she found a new phage (a virus that replicates and infects within bacteria) called Lambda (λ) lurking inside a bacterium that was supposed to resist explosive phages. However, when Lederberg mixed two bacteria, the virus-resistant form and a virus-sensitive strain, the mixture produced telltale signs of the lambda phage. The resistant bacteria harbored an inactive phage λ. But when mixed with other bacteria, the sleepy phage emerged and killed the other bacteria, revealing its presence for the first time.

It was a significant discovery because Lederberg and others could then easily study genetics. Lederberg was interested in how the phage and bacteria interacted. During her experiments, she studied the famous Fertility Factor called F.

She used genetic methods in the laboratory and found short segments of DNA harboring genes for transferring between phages and bacteria. Lederberg discovered that these DNA pieces coded for the Fertility Factor. The DNA transfer process became known as genetic recombination. The Fertility Factor and its genetic movement between viruses and bacteria could be used to make a genetic map, a blueprint showing where the various genes are located on the genome.

Lederberg was also an inventor. During graduate school, she developed a method for copying bacteria on Petri dishes and using the copies to map genes and trace their recombination events. Lederberg called this invention "replica plating" because the Petri plates were "replicates," i.e., copies of original dishes. We read about Petri plates and dishes in Chapter 2. Soon geneticists were producing genetic maps for other viruses and bacteria. Later, a human genetic map was developed using the same sort of genetic approaches that Lederberg had used.

Thanks to the pioneering efforts of Esther Lederberg, we now have a clear understanding of the nature of genes and genomes and how they recombine to produce new variants. Today, it is possible to map a person's genes and possibly find their ancestors or whether they might suffer from a genetic disease inherited from their parents and grandparents.

CHAPTER 14

YVONNE BRILL

1924–2013

ROCKET SCIENCE

Spaceships, rockets, and other spacecraft that explore outer space need propulsion. Dr. Yvonne Brill was able to help to develop a method to provide such propulsion.

Brill was born in Manitoba, Canada. She endured much discrimination as a woman attempting to succeed in a male-dominated world. Brill studied math and chemistry and graduated from college at the age of 20. The University of Southern California later conferred her master's degree in chemistry.

She studied ammonia, hydrogen, and nitrogen and attempted to produce a more fuel-efficient chemical propulsion system for rockets. Mixing these elements, Brill made a new rocket fuel called "hydrazine."

One of her amazing intellectual insights was to connect an electric heater to a high air-intake jet engine called a ramjet. She invented a new rocket engine called an Electrothermal Hydrazine Thruster (EHT/Resistojet) which she patented. It was a remarkable advance in the growing field of rocketry! Brill's thruster device was revolutionary, permitting rocket scientists to send hundreds of rockets carrying satellites into orbit around Earth.

Over the many years in which she worked, Brill contributed to the following:

- The Nova project that took American-made probes to the moon
- The first weather satellite propulsion system
- The first satellite placed in the upper atmosphere
- The Mars Observer and the engine for the space shuttle

Due to her pioneering work, satellites can remain in orbit for long periods. Thanks to Brill's remarkable rocket propulsion innovations, we now enjoy rapid electronic communications worldwide and accurate weather predictions. We also know a great deal about planetary science.

CHAPTER 15

IRINA PETROVNA BELETSKAYA

1933–

ORGANIC CHEMIST

Dr. Irina Petrovna Beletskaya is best known for her studies of chemical bonds between carbon atoms and metals, such as nickel and platinum. She was trained in graduate school in Russia, took her doctorate at Lomonosov Moscow State University, and ultimately became a professor at the same institution. She is a remarkable scientist who has made many contributions in the field of organic chemistry.

One challenge Beletskaya encountered during her lifetime was refusing to join Russia's Communist Party which during the Soviet era, was the only political party in what was then the Soviet Union. As a result, she was barred from being a lab leader and had to deal with political peer pressure.

Beletskaya was very good and insightful regarding organic chemistry, the study of carbon-containing chemicals. Carbon is a common atom that likes to make four bonds with other atomic particles, such as hydrogen, and forms the basis of life on Earth. All living beings are said to be carbon-based life forms.

Organic chemistry studies carbon, hydrogen, and oxygen atoms inside molecules. Beletskaya became a leading expert on the bonds or links between carbon and metals. She discovered that metals make specific organic chemicals react better than they would without them. Making chemical reactions happen more powerfully is called a "catalytic" behavior. Beletskaya found that metals make great catalysts. She used these metal catalysts to study how chemical bonds formed and their bond features.

Beletskaya was a pioneer in starting a new scientific subject called "green chemistry." The idea behind green chemistry is to make things people need without generating much waste as they manufacture them. Beletskaya was thinking of things such as medicines, industrial materials, and electronics. She found that her metal catalysts were beneficial in assisting green chemistry.

Another area of interest to Dr. Beletskaya was chemical weapons, which can cause serious injuries to people or even kill them when such agents are used. Beletskaya's primary intent was to invent new ways to neutralize old and outdated chemical weapons without too much effort. She focused on agents such as mustard gases, which cause serve burns on the skin. Beletskaya's scientific efforts provided new avenues for getting rid of expired chemical weapons. Thanks to Dr. Beletskaya, the world is better, cleaner, and safer.

CHAPTER 16

MARY GAILLARD

1939–

ATOMIC PHYSICS

Dr. Mary Gaillard was known for her amazing work in "theoretical physics." She studied at the University of Paris in Paris, France, and developed many theories regarding contemporary physics.

A hypothesis in science is a hunch, a guess, a presumption, an idea, or a belief about something. Until we can support the idea with scientific evidence, a hypothesis may be believed or acted upon, but a theory has the weight of scientific confirmation—it is frequently supported by data. Of course, people can have opinions about things but that does not mean they are correct!

Gaillard's experience in getting her Ph.D. was difficult. This was not uncommon for women studying science, in part, because there were so few of them. She was a trailblazer in the field of women studying physics. Her main advisor had died, and she was often discouraged by others for wanting to study physics. Some physics laboratories even barred women from doing experiments or had quotas, allowing only a few women to work in labs. She overcame these obstacles and made significant contributions to theoretical physics.

Gaillard was interested in the substances that make up the universe. We know that all matter in the universe is composed of very tiny atoms. In general, physicists and chemists understand that atomic elements have protons, electrons, and neutrons. Protons and electrons have opposite charges, while neutrons have no electrical charges.

Gaillard also studied "quarks." She wanted to understand what was inside of atoms, stuff that scientists called "particles." Gaillard learned that these so-called sub-atomic particles had forces within them holding the universe together. She designed a theory to explain particle physics called the Standard Model, which described the elemental energies and the building blocks of the universe's mass. It was an original and ingenious theory!

Gaillard's Standard Model predicted the presence of sub-atomic pieces with great names, like the charm quark and the strange quark, or less interesting designations, like the b-quark. Gaillard was an expert in these particles, and in graduate school, she discovered one she called the "kaon."

What is a "quark?" They are very tiny electrically-charged pieces inside the non-charged neutrons of atoms. Neutrons have no electric charge, but their insides have two charged quarks Each quark type is called "up" or "down." Eventually, many other quarks were discovered. All matter that can be observed is composed of quarks. Scientists are still studying the quarks.

Gaillard made significant contributions to other fields of theoretical physics. One famous discovery is "supersymmetry." The topic describes how each atomic particle has an opposite partner, called a "superpartner," with unique properties like their energies, masses, and charges. It was all very new and exciting!

Gaillard was interested in a theory called "supergravity." She combined Albert Einstein's famous theory of general relativity with supersymmetry. It was all quite informative to physicists and legendary for its insight.

Another valuable contribution by Mary Gaillard was her studies of what is known as superstring theory. She described the insides of atoms as behaving like wiggling strings! Despite getting less pay for the same amount and quality of work than men, Gaillard broke many boundaries and paved the way for girls and women to study science and work in scientific fields at the highest levels.

CHAPTER 17

JOCELYN BELL BURNELL

1943–

PULSARS, SIGNALS FROM OUTER SPACE

Dr. Jocelyn Bell Burnell became famous in astronomy, the study of outer space, for being the first person to discover the pulsar. These pulsars are radio stars that quickly spin while emitting a beam of radiation that Jocelyn Bell was the first to find—it was a historic achievement and she made it while she was still in her 20s!

As a student in graduate school, Jocelyn built a massive radio antenna to find quasars. These outer space structures are supermassive black holes in the centers of galaxies. Bell operated the device and collected the data from outer space. She looked for quasars, hoping to find new ones.

Instead, Joelyn noticed that her radio antenna showed tiny blips on the ink tracings on large reams of paper with a special pattern. The blip patterns, which Bell called "scruff," were not from Earth!

Nor were the signals from anything made by humans in space, like Earth-orbiting satellites. But the blips were constant, a property she called periodic, occurring every 1.3 seconds!

Jocelyn even took pains to rule out aliens from somewhere in the galaxy. Bell called these periodic signals "pulsars" because they pulsed in energy signals on her radio antenna. She ruled out aliens as a source of the signals from outer space as they did not come from a planet orbiting a sun-like star, similar to our Earth's sun. Yet, just for fun, Bell named the first pulsar "LGM-1" for "Little Green Men!"

Bell's discovery of the first pulsar to be observed was the focus of her doctoral thesis project. She received her Ph.D. in radio astronomy from the University of Cambridge. Bell did not receive the Nobel Prize for her discovery. Instead, her advisor did, and many think this is unfair.

"Bell's Star," as her pulsar later became known, is 978 light years away, meaning that if one traveled to it at the very fast speed of light, it would take many years to arrive there! Today, we know that Bell's Star is a white dwarf. These stars were once massive neutron stars that went "supernova," experiencing a super enormous explosion and then shrinking to a tiny shell of itself and emitting a turning pulse, like a beacon.

Speaking of beacons, Jocelyn also studied radio galaxies. These supergiant structures have massive centers that emit radio waves that radio antennae, like Bell's, can readily detect. She became a world expert on these radio galaxies. These giant structures in outer space exist extremely far away outside of our own Milky Way galaxy. She is one of the great astrophysicists in the history of science. By any standard, her work and discoveries are remarkable and truly out of this world!

CHAPTER 18

SUSAN GOTTESMAN

1945–

GENES AND PROTEINS

Dr. Susan Gottesman went to Radcliffe College, a widely-admired institution, where she concentrated on biochemical sciences. Like Marie Daly before her, Gottesman read the classic 1926 book Microbe Hunters by Paul de Kruif. It as an extremely influential book among scientists. For fun, Gottesman studied microbes that turned bright red when grown on food agar in Petri plates. The red spots on the agar Petri dish are colonies of the bacteria. And scientists learn a lot from studying living beings in these Petri dishes.

After graduating from college with honors, Gottesman investigated microbes at Harvard University for her Ph.D. in microbiology. She studied in Harvard's Department of Bacteriology and Immunology. The scientists there studied microbes and the immune response.

In graduate school, Gottesman's first major scientific study involved turning on genes coding sugar- and amino acid-eating behaviors in bacteria. Amino acids are building blocks of protein. Eating amino acids and sugars allow bacteria to get the energy to grow as we humans do. Gottesman discovered that these food-eating genes could move from bacteria to bacteriophage viruses that infected them.

Gottesman took advantage of these microbe gene movements to create new bacterial mutants. One of these mutant groups controlled genes by making loops out of the DNA that coded the sugar-eating behaviors! It was very exciting to control genes by looping DNA strings. Gottesman became famous among fellow scientists for her gene-controlling studies.

Working as an independent microbiologist and scientist, she focused on discovering how bacteria destroy their protein. Gottesman discovered a new gene that controlled protein degradation. In an elegant study, she killed the gene, and the bacteria lost their ability to destroy the protein. It was a fantastic achievement.

Gottesman discovered a new way to regulate protein making as an established scientist. Her discovery involved using small RNA molecules to do so. These short RNA molecules bind proteins which then control their protein activities.

This research project involved another fascinating microbial machine. Gottesman discovered that RNA made "hairpin" shaped loops. Gottesman's small RNAs formed structures that look like hairpins that people use to arrange their hair. Bacteria use these RNA hairpins to control how proteins work.

Gottesman's investigations of genes and proteins are greatly valued among scientists. Because of her contributions, we know much about how bacteria deal with stressful conditions—they regulate specific genes and proteins to help them survive stress. Given the amount of stress we are all under, this is important!

CHAPTER 19

SHOBHONA SHARMA

1953–

MALARIA

Dr. Shobhona Sharma was born and raised in India. She was trained in chemistry, taking her bachelor's and master's degrees in Mumbai, one of the largest cities in the world. Her doctoral thesis was focused on learning about plant enzymes.

After earning her doctorate, Dr. Sharma moved to New York City and studied disease-causing microbes. One of these microbes causes the terrible disease of malaria. Sharma researched the relationship between parasites (which often suck the blood from people) and malaria.

What exactly *is* malaria? The disease is characterized by fever and extreme tiredness. It is caused by the bite of a specific mosquito that spits its saliva with tiny worm-like microbes called *Plasmodium* into our bodies while the mosquito takes a blood meal. Before we knew this, people thought the condition was caused by "something terrible in the air," hence, the name malaria, which has its roots in the Medieval Italian phrase *mal aria,* meaning "bad air."

The malarial microbe (*Plasmodium*) is a tiny worm that lives inside our red blood cells causing terrible high fevers and extreme tiredness among its victims. Until modern times, we did not know how malaria was transmitted to people. We know now that mosquito bites can cause infection when they spit into our blood while injecting the tiny forms of the worms.

We probably have all heard the word immunology—what exactly does it mean? The immune system can protect healthy people from dangerous microbes by making antibodies. When the microbes invade the individual again, antibodies quickly destroy the microbe.

Sharma found a link between the mosquito and many diseases. What was that link? She learned how the body's immune system could protect people from malaria. These antibodies are created in response to the invading microbes. Sharma discovered a protein in the malarial worm that could be used as a target to generate antibodies by the immune system.

A parasite is a creature that has difficulty growing by itself. However, a parasite can grow better if it steals living cell parts from another being. Unfortunately, stealing things from living cells can be dangerous to the cells and the living beings attached to them because they, too, need those cells to be healthy.

Studying parasites is important because they can cause serious disease outbreaks in unsuspecting victims. Suppose we can understand their structures, how they work, and how the immune system responds to them. In such instances we can find new ways to avoid them, treat them with medicines, or prevent disease by vaccination.

Sharma also studied cancer, a disease that can be debilitating and deadly. Our body's immune system can help in fighting cancer. Sharma studied how when specific white blood cells turned cancerous, a condition called lymphoma, their genetic sequences were limited in their scope. This finding suggested that lymphoma had fewer targets than thought before.

Additionally, Shobhona studied specific viruses that caused cancer. She made a big discovery about it. Dr. Sharma found a connection or correlation between some viral genes and their effects on immune cells. This new connection made it possible to develop antibody-making plasma cells.

Dr. Sharma has enjoyed a very productive and rewarding career as a scientist. Thanks to her efforts, we can better protect ourselves from microbes and cancer.

CHAPTER 20

KATALIN KARIKÓ

1955–

RNA BIOLOGY AND COVID-19 VACCINES

Dr. Katalin Karikó is an American-Hungarian and amazing scientist whose research has benefited all of us who were vaccinated during the COVID-19 pandemic. When a disease affects people from multiple countries or continents, it is considered a pandemic, a worldwide disease occurrence. Even though the COVID-19 pandemic has taken countless lives in a short time, Karikó's work has saved many, perhaps even billions, of lives.

Karikó was born in Hungary and took her Ph.D. from the University of Szeged. Afterward, she moved to the U.S., where she took up new research on several fronts. However, she encountered trouble getting grant money to do her research. Nevertheless, she persisted and eventually was successful in many scientific ventures.

Karikó became an expert on the biochemistry of RNA, which has to do with making proteins. She was interested in using RNA molecules to treat human diseases. She investigated the chemical structure of RNA called ribosomal RNA (rRNA), part of the protein-making machinery of everyone's cells. Karikó found that the modified rRNA was safe from degrading enzymes after she altered its structure.

She also studied the immune system and its relationship to rRNA. Karikó found that rRNA provoked a very weak immune response. This new finding was a significant one, and it made Karikó a famous scientist.

Another type of RNA is called messenger RNA (mRNA). Karikó's discovery regarding rRNA would prove crucial for developing an mRNA vaccine for COVID-19. Normally, mRNA has the genetic code in it for making specific proteins with their proper sequence of amino acids that make them up. Scientists thought that if they injected mRNA coding COVID-19 spike proteins into patients, it would induce the cells to make the spikes. Then, the body could make antibodies to target the spikes, making a vaccinated person immune to infection by the SARS-CoV-2, the virus that causes COVID-19.

The mRNA vaccine coded only for spike proteins and *not* the entire virus. Thus, the vaccine could not cause COVID-19 disease but it could provoke the immune system. If such a vaccinated person gets infected with SARS-CoV-2, the body would recognize their spikes and respond to them with neutralizing antibodies, reducing the severity of the disease. Such vaccinated people likely would not need hospitalization and could survive the otherwise deadly disease.

Regular mRNA had a massive problem—it's weak. If mRNA is left out in the open, it soon gets destroyed by specific enzymes in the environment. Karikó considered the RNA stability problem. To make an effective mRNA vaccine, scientists needed to make it tough. Karikó's expertise in making rRNA resilient to destroyer enzymes was a perfect solution. She and her colleagues made the spike mRNA tough and strong by modifying it to behave more like rRNA.

The modified mRNA was thus made more resilient using Karikó's knowledge of RNA chemistry, and effective RNA vaccines against COVID-19 were made and widely used. For this astonishing achievement, Karikó received the 2023 Nobel Prize in Medicine. In the future, RNA can become even more important for preparing vaccines for other disease-causing microbes and treating genetic diseases. In part, we can thank Dr. Karikó for that!

CHAPTER 21

DOROTHY CROWFOOT HODGKIN

1910–1994

PENICILLIN AND INSULIN STRUCTURES

Dr. Dorothy Crowfoot Hodgkin is a famous biochemist and Nobel Prize Laureate in the field of chemistry. She is best known for her studies of the structures of penicillin and insulin, both very important molecules in treating people with infections and diseases such diabetes.

Dorothy was born in 1910, on May 12, in Cairo, Egypt. Her parents were John Winter and Grace Mary Crowfoot. Dorothy was raised in England and became fascinated with chemistry at age eight. She grew up during World War I (1914–1918), and was educated in the country, away from larger schools where science and math were taught.

She was enrolled in various small schools where girls were not taught science and math. She was dismayed that she was not qualified for college at the University of Oxford because she did not have courses in Latin and the sciences. So Hodgkin learned Latin and was tutored in the sciences by her mother. Thus, although Dorothy was behind in these areas, she quickly learned chemistry and Latin. Another problem for her were finances—Dorothy had no money for college tuition. Fortunately, her doting aunt, Dorothy Hood, provided the required funds for tuition.

Overcoming these obstacles in her education and funding, she was enrolled at Somerville College at the University of Oxford in 1928. Hodgkin studied chemistry and a technique called X-ray crystallography. She graduated with her bachelor's degree in 1932. Hodgkin's next venture was attending the University of Cambridge to earn a Ph.D.

An exciting story is told about Hodgkin's path to graduate school. The life-changing incident occurred when her father was on a train and met up with an old friend, Thomas Martin Lowry, who suggested that Dorothy might consider working for the famous scientist John Desmond Bernal, a pioneering X-ray crystallographer. With additional funding from her Aunt Dorothy, young Hodgkin would study the structures of cholesterol and pepsin in Bernal's laboratory at the University of Cambridge.

Cholesterol is an important molecule because it helps the membranes of cells and plays a role in heart disease. Pepsin is an enzyme necessary for digesting nutrients. For her thesis project, the crystal structures of cholesterol and pepsin were determined by Hodgkin, who received her doctoral degree in 1937.

In 1946, Dr. Hodgkin used her expertise in X-ray crystallography to study the structure of penicillin, a vital medicine for the treatment of infection. At the time, penicillin was a widely-prescribed antimicrobial molecule. Hodgkin made crystals of these important molecules and zapped them with X-rays, which bounced off the crystals, a process called diffraction. Hodgkin then used her data from the

X-ray diffracting angles to construct contour maps showing the electron densities. Using this method of analysis, Hodgkin determined the shapes of the molecules.

When Hodgkin solved the question of the structure of penicillin she discovered that it had a special chemical arrangement called the beta-lactam ring (i.e., β-lactam), which was the active ingredient of the β-lactam molecules. If the β-lactam ring was intact, that penicillin could work against infection, and if it were not intact, it would not work in treating infection. This was an extremely important finding.

Another study by Hodgkin involved the structure of insulin. Compared to penicillin, insulin was a very large molecule. Insulin is known to harbor over 700 individual atoms and two large protein chains called insulin subunits A and B. Insulin is a large molecule. It has a molecular weight of about 5000 Daltons. Solving its structure would be a major undertaking.

As she did earlier, Hodgkin exposed the crystal structures of insulin to the X-rays, which diffracted, and like before, she made a contour map of the insulin structure using the diffraction angles. When she finished, she published the discovery in *Nature*, a prestigious scientific journal, and in 1969 she became very famous.

Vitamin B-12 (cobalamin) is an important vitamin as it is necessary as a cofactor for enzymes to mediate their chemistry. Without vitamin B-12, people can suffer detrimental effects such as pernicious anemia, a debilitating condition. Hodgkin solved the puzzle of the Vitamin B-12 structure in the late 1940s.

In 1964 Dr. Hodgkin earned the Nobel Prize in Chemistry for her structure studies of penicillin cholesterol, insulin, and vitamin B-12. Only two women before her had received the Nobel Prize, Drs. Marie Sklodowska Curie and Irène Joliot-Curie, her daughter.

CHAPTER 22

BARBARA McCLINTOCK

1902–1992

GENES THAT JUMP!

Dr. Barbara McClintock is one of the most famous female scientists who has ever lived. She is responsible for the famous discovery of genes that jump. She discovered that not all genes on a chromosome remain in place—they move. These jumping elements are called transposons. The transposons are genetic elements that move from one location to another.

Barbara was born in Hartford, Connecticut in 1902. Her parents, Thomas and Sarah McClintock, raised her in Brooklyn, New York. Barbara was an energetic youth who liked sports such as swimming, skating, and volleyball.

She graduated high school at age 17 in 1919 and went to New York State College of Agriculture, located at Cornell University. While in college, McClintock played in a jazz band and became the class president of the first-year students. In 1923, she received her bachelor's degree in agriculture, concentrating on botany and plant breeding. She later received a master's degree in botany. In 1927, McClintock finished her Ph.D. in the botany department at Cornell.

McClintock was a pioneer geneticist who studied the genetics of corn with the scientific name *Zea mays*. She was also notable for her discovery of genetic recombination and the crossing over of genes in corn plants.

In graduate school, McClintock learned various cultivation techniques and plant breeding with corn. She learned how to cross-pollinate corn and grow them to study corn genetics. She learned how to purify corn DNA and stain them so that she could see the chromosomes under the microscope.

McClintock would harvest the corn, collect their anthers (the part of a stamen that contains the pollen), remove their cell walls and their flower parts, squeeze the anthers' contents onto a microscope slide, and then add a chemical stain to the chromosomes. She could add a cover slip and flame the stained corn chromosome mixture to fix the corn chromosomes to the glass slides. Next, she added an acetic acid chemical, and the coverslips would come off, leaving behind the chromosomes stuck to the microscope slides. Then, she washed the glass slides with stained corn chromosomes. It was an arduous, painstaking process. It was worth the effort because McClintock could see the stained chromosomes under the microscope.

After earning her Ph.D., Dr. Barbara McClintock collaborated with a graduate student named Harriet Creighton and studied genetic recombination and gene crossover. The two young scientists used a set of gene markers that encoded knobs, which were visible at the ends of the corn chromosomes.

The genes that McClintock and Creighton studied were linked. These genes were close together along the corn chromosomes. They used genetic markers, like those coding for pigmentation in corn protein called aleurone, colorless aleurone, waxy kernel starch, and shrunken endosperms—the starch and protein storage compartments—to study the movement of genes as they crossed over between and within corn chromosomes.

Their work was arduous. Creighton and McClintock worked in experimental cornfield stations from sunup to sundown. They planted the kernels with their observable characteristics in the spring, watered, and weeded the growing corn plants in the hot sun while maintaining records of every corn plant, including their genetic histories.

They had no control over the weather, drought, or rain. They would lose all their efforts if their corn plants failed to grow. After the harvest of the corn, they undertook their painstaking work to study the chromosomes under the microscope. In so doing, Creighton and McClintock had observed genes that were moving to new locations on the chromosomes. It was the first time in scientific history that genetic recombination had been seen during the cell division process called meiosis, in which chromosomes are separated into germ cells called gametes, like eggs and sperm.

In 1950, Barbara studied how chromosomes break and fuse in corn plants. This work led to her famous discovery of the so-called transposons, i.e., the famous jumping genes. A particular chromosome location, called a mutable locus, had a higher rate of chromosome breakage than normal. McClintock named this gene region on the chromosomes that was frequently broken as "D" for dissociation.

Interestingly, McClintock's work on transposition (gene jumping) was met with great disbelief amongst other scientists, and it would be decades before she was proven correct and given credit for her historical work. McClintock would not be awarded the Nobel Prize for her discovery until she was in her 80s, over 30 years after the first discovery of the jumping genes.

In later years, McClintock would become a key scientist in the studies of the origin of modern corn. She was an important scientific figure in solving the origins of corn genomes and how ancient corn evolved into our now-modern corn that much of the world uses for food. Other scientists discovered the molecular process of how the genes jumped. Corn uses transposase enzymes that would cut the DNA at specific locations. Then, these enzymes move the cut DNA parts to new locations and insert them into specific insertion sites called motifs located in other locations around the genomes.

In contemporary science, the work is still important, and we can now study genetic recombination that occurs from generation to generation. Dr. McClintock's studies played an important role in these genetic efforts.

CHAPTER 23

GERTRUDE ELION

1918–1999

BIOCHEMIST AND BIOMEDICAL SCIENTIST EXTRAORDINAIRE

Gertrude Belle Elion was a famous biochemist who made many great contributions to biomedical science by developing new medical treatments. At age 70, Elion would become a Nobel Prize Laureate for her discoveries involving new strategies for chemotherapy in the treatment of several important diseases.

"Trudy" Elion was born in the Bronx, New York, on January 23, 1918. Her parents were Robert and Bertha Cohen Elion, both of whom were of Jewish ancestry. Elion's father, a Lithuanian immigrant, was a dentist, and her mother was a Russian immigrant, having moved from what is now Poland.

Historians have described the Elion household as an environment conducive to intellectual pursuits. The family partook in voracious reading activities. They were devoted to studying a diverse array of topics, including history, classical literature, biographies, fiction, and poetry.

As a child, Elion was considered exceptionally bright compared to her fellow students. She regularly skipped grades as she progressed through elementary and junior high school. Thus, she graduated early from Walton High School in New York and, at 15 years old, was extremely interested in attending college to study chemistry.

However, during the stock market crash in 1929, funds for college became a problem. Because tuition was free, Elion enrolled in Hunter College, an all-women college. She graduated in 1937 at 19 years of age in chemistry with honors, such as being inducted into the honor society *Phi Beta Kappa* and graduating *Summa cum laude* ("with highest honors"). Despite her strong academic performance in a challenging field of study, Elion was rejected by every graduate school to which she had applied, each school telling her that no funds were available to her for any graduate programs.

Elion also experienced tremendous difficulty in finding a job as a chemist. Therefore, she was encouraged to go to a secretarial school, which she did, and it was incredibly discouraging. Nevertheless, she managed to save enough money to enroll in graduate school and pay for it independently. It was a tremendous achievement, especially since male graduate students did not encounter such financial hardships.

Elion enrolled in a master's program at New York University and taught chemistry and physics part-time at the high school level in the New York public school system. In 1941, Elion earned her master's degree in chemistry from New York University.

During World War II, there was a shortage of male chemists and Elion managed to secure a quality control inspector job with a supermarket and later with a food company analyzing various foods such as mayonnaise for color and pickles for their acidic quality.

Elion found a new job working for the pharmaceutical company Johnson & Johnson as a laboratory assistant in developing antimicrobial drugs like sulfa drugs or sulfonamides. Unfortunately, a new vice president of the company changed directions, and Elion's work was changed to less satisfying endeavors. In 1944, she had an opportunity to work under a prestigious professor, Dr. George Hitchings, who was in charge of a research section at Burroughs Wellcome Research Laboratories. Prof. Hitchings would become a mentor and an avid supporter of her work. In 1988, they shared a Nobel Prize in medicine for inventing innovative approaches to designing novel medicines.

In her new position as a laboratory chemist, she studied one of the nucleic acids called purine using sophisticated laboratory equipment, such as spectrophotometry. The instrument could measure the amount of light absorbed by substances or transmitted through them. The machine could trace the progress of chemically making new materials. Elion successfully synthesized new nucleic acid derivatives and found the work satisfying.

Her success in this area inspired her to pursue a doctorate. However, she would never earn the coveted Ph.D., though many years later, she was given many honorary doctorates for her astonishing scientific successes. She enrolled in graduate school at the Brooklyn Polytechnic Institute in a Ph.D. program. However, she was not supported in her efforts as a Ph.D. student. The college dean discouraged her from graduate school by informing her that she could not simultaneously work *and* be a graduate student. One had to be given up. So she gave up graduate school to continue working.

Elion's first major scientific achievement involved leukemia, a disease that children can suffer from, involving a cancer of white blood cells in which they grow in larger than normal numbers. She found that a compound known as 6-mercaptopurine was a potential anti-leukemia medicine, initially showing good efficacy. However, the effect seemed to wane after a while and did not work so well against leukemia. The failure of 6-mercaptopurine to treat leukemia long-term was a major disappointment.

She learned that the 6-mercaptopurine was broken down into thiouric acid using an enzyme called xanthine oxidase. It was a major discovery in biochemistry. Elion reasoned that to make her 6-mercaptopurine agent work better, an inhibitor of the xanthine oxidase would leave 6-mercaptopurine alone. That was when she tried a compound known as allopurinol. It blocked the enzyme, but it took so much of the allopurinol that it proved too toxic for leukemia patients. It was another disappointment.

Allopurinol would later become a famous molecule because of a terrible condition called gout. Gout is a painful disease characterized by arthritis in the joints. Blood levels of uric acid are overproduced and oxidized by purine metabolism, leading to the deposition of crystals containing uric acid in the joints and the painful symptoms of gout. Gout symptoms include inflammation of the arthritic joints, pain, tenderness, redness, heat, and swelling of the joints.

Patients with gout often experience higher than normal uric acid levels in the blood, known as hyperuricemia. Elion reasoned that if allopurinol was, in fact, effective in inhibiting the enzyme, then it might lower the toxic levels of uric acid in the blood. Elion discovered that allopurinol was important in improving the condition of gout by lowering the excess uric acid concentration in the blood. It was a major finding.

Thanks to Elion, gout treatment involves allopurinol. The U.S. Food and Drug Administration (FDA) approved allopurinol for the treatment of gout in 1966. She became very famous for this discovery. Trudy also was involved in other discoveries. For instance, she participated in the discovery of an anti-viral agent called acyclovir, which is effective against infectious diseases like herpes, cancer, and HIV infection.

Elion was also an important scientist in discovering an anti-bacterial agent called trimethoprim. The agent targeted metabolic enzymes called dihydrofolate reductase and xanthine oxidase, both involved in nucleic acid biosynthesis. These enzymes are essential intermediaries in metabolism to produce amino acids for protein and nucleic acids for DNA and RNA. These building blocks are essential to life.

Although Gertrude Elion never acquired the coveted Ph.D., she was routinely honored with honorary degrees. The fact that Elion has managed to make so many extraordinary contributions to science is a true testament to her resilience.

As a highly-successful scientist, working in work climates that were not conducive to women or persons of Jewish origins, Gertrude "Trudy" Elion rose above many obstacles and became a truly successful scientist.

CHAPTER 24

KATHERINE GOBLE JOHNSON

1918–2020

HUMAN CALCULATOR

Katherine Johnson was an extraordinary African-American woman who overcame adversity to succeed in a world dominated predominantly by men. Her work in mathematics and what is known as orbital mechanics was critical to the U.S. space program, especially for early spaceflights by astronauts. Her life was astonishing from the start. For instance, she began high school at age ten and graduated from college at age eighteen. She had to move to a city that admitted black students in its schools, abide by segregated bussing rules, and uphold excellent grades to accomplish such an achievement. Her intellect allowed her opportunities atypical of her classmates. All the while, she withstood racism and faced it unabashedly.

Katherine was the fourth child welcomed into the home of Joshua and Joylette Coleman. On August 26, 1918, Johnson's father worked as a lumberman, farmer, and repairperson, while her mother was an educator. She attended public school until the eighth grade, the highest level of education offered at that time in Greenbrier County, West Virginia. Her parents registered Katherine and her siblings in high school at the Institute, West Virginia. Although Katherine was only ten years old when she started high school, she excelled and began her college career at the age of fourteen at West Virginia State College. The founding of West Virginia State College provided higher learning gateways for black students; however, students of all races could obtain admittance. Selection as the first African-American to attend West Virginia University was an honor given to Johnson in 1939. Many of Johnson's professors mentored her and provided unique mathematics courses expressly for Johnson. By the time she was eighteen, Johnson had received her undergraduate degree in Mathematics and French.

After graduation, Johnson remained in Virginia and taught at a black public school. Within two years, she married her first husband, James Goble, left her teaching job, and began graduate school with an emphasis on mathematics. Johnson became the first African-American woman admitted to graduate school at West Virginia University. After just one year, however, she decided to give her full attention to her family as she learned of her pregnancy.

She joined the National Aeronautics and Space Administration (NASA) group in 1958. Her precision with mathematical computations was legendary. Astronaut John Glenn reportedly trusted her calculations over those of a computer. Johnson first became involved with the National Advisory Committee for Aeronautics, known as NACA, in 1958 after the Soviet satellite Sputnik was launched into space before the U.S. had launched a satellite and starting what was called the "race for space." Johnson had provided some mathematical calculations for the 1958 document *Notes on Space Technology*, assembled from a series of lectures by engineers in the Flight Research Division and the Pilotless Aircraft Research Division.

Johnson's conduct and endeavors toward equality made it conceivable for other women at that time to receive better treatment in the workplace. She was a voice of reason for equal rights among the races. Johnson inspired her co-workers and others to treat each other with self-respect and admiration. Modestly, she often claimed that she was just doing her job.

Johnson dubbed herself and the other female employees at NACA (later known as NASA), as "human computers." She and the other women read the data from the aircraft's black boxes. The data indicated specific mathematical tasks and their interpretation of it enabled NASA to take the lead in the space race.

Because of her understanding of analytic geometry, Johnson's work involved assisting an all-male flight research team. Her assertiveness gained her access to briefings, which were traditionally off-limits to women; these meetings, meant to be a think-tank for engineers to plan a space mission, did not welcome women. She declared that the computational work was hers and justified attending the editorial meetings.

Katherine used geometry when calculating the height and speed of the parabolic flight path for the first American human-crewed mission in space, piloted by Alan Shepard. She said the early trajectory was parabolic in shape, and predictions about where it would be at a given time would be reasonably easy to determine.

By 1962, NASA used electronic data processors to calculate launch conditions for the Friendship 7 mission. The calculations necessary for orbiting the Earth were definitely more complicated than the ones needed for the first mission in space. Its astronaut, John Glenn, trustingly insisted that Johnson check over the computer's calculations for accuracy.

Additionally, Johnson published her trajectory analysis, along with Ted Skopinski's contributions, in a report detailing the equations relating to an orbital spaceflight, which specified the splashdown site of the returning space capsule. She also contributed to the orbital mission of John Glenn, the first American to orbit Earth and the first human mission to land on the moon. Johnson's computations influenced every major space program beginning with Project Mercury, the first human spaceflight program, through the Space Shuttle program. When interviewed about her space exploration contributions, she remarked that her most significant contribution was determining the necessary synchronization procedures for Project Apollo's Lunar Lander with the moon-orbiting Command and Service Module.

Johnson used mathematical equations dating back to English scientists Sir Isaac Newton (1643–1727) and Henry Cavendish (1731–1810). More specifically, the laws of gravity afforded Johnson the necessary data to compute the force of gravity between the Earth and the spacecraft. Spacecraft in low orbits travel at higher speeds because the gravitational pull is intense. In higher orbits, spacecraft travel at lower rates because the force of gravity diminishes.

Johnson co-authored twenty-six scientific papers throughout and about her NASA career. These publications resulted from her attendance and participation in the NASA consultations. She wrote about her calculations concerning spacecraft trajectory, orbital behavior, optical measurements, nonlinear aircraft maneuvers, and space antennae. Though it was slow in coming, she eventually gained notoriety and respect for her efforts.

Johnson's employment at NASA broke race and gender barriers. Although she initially worked in a team of African-American women, she ultimately worked in a mixed-gender and mixed-race environment. Something as rudimentary as using the restroom also confronted Johnson and the other African-American women at that time. Segregation existed for cafeterias and lavatories. Johnson's office sign read "Colored Computers."

The inclusion of a woman's name on a report at that time at NASA was unprecedented. That might sound ludicrous in the twenty-first century or at any time, but it was standard operating procedure (SOP) back then. Johnson changed that SOP when she had her name contained within a report she contributed and finished writing. She stated the necessity for assertiveness and aggressiveness on the job at varying degrees, depending on one's desired goal.

Katherine Johnson received numerous awards and honors, such as the Medal of Honor from the Daughters of the American Revolution. In Fairmont, West Virginia, there is a building named the Katherine Johnson Independent Verification and Validation Facility. She received several honorary degrees from various universities, such as a Doctor of Law, a Doctor of Science, and a Doctor of Humane Letters. Johnson even received the Silver Snoopy award for her exceptional contributions to flight safety and mission success.

Katherine received the Presidential Medal of Freedom in 2015, the nation's top civilian honor, which she proclaimed her most prized honor. Johnson dedicated her life to mathematical and scientific endeavors. Her teaching career and time at NASA provided her love of learning an opportunity to mature and flourish.

After her retirement from NASA, Johnson spent time with her family, specifically her three daughters, and traveling. She also devoted some of her time to the lecture circuit, speaking to students about studying math and science and working toward their aspirations and goals.

The 2016 movie *Hidden Figures* describes Johnson's experiences during her NASA employment and the barriers faced by African-American women during that era. Katherine Johnson was an exceptional person who was a trailblazer in America's space program. Her inspirational life and work was a powerful example for any person.

AFTERWORD

Each of the women in this book made contributions to science and medicine that are remarkable. Because of their effort, we know more about many scientific disciplines.

Often, their work was made more difficult because of the times in which they lived—an era when there were fewer opportunities for women who were interested in science than there are today. But these women did much to advance not only science, but the opportunities for others who want to do the same. You can follow in their footsteps!

If you want to look through a telescope or a microscope; study the stars, the seas, or the cells and all that sustains life; you can do it!

There is much work to be done in the sciences and many discoveries yet to be made—made by someone like you!

Glossary

Agar is a jellylike material isolated from different red seaweed types and used as a food thickener and in biological labs for culture media. Angelina Fanny Hesse was the earliest person to suggest using agar to grow microbes.

Antigens are molecules foreign or native to the body and can bind specifically to an antibody or generate pieces recognized by specific white blood cells with dedicated T-cell receptors.

Atoms are the fundamental parts of all matter, consisting of nuclei, protons, electrons, and an array of sub-atomic particles.

Bacteria are unicellular microorganisms called prokaryotes that do not have organelles or a nuclear membrane. While most bacteria are not harmful, some can cause disease.

Chaulmoogra oil is a thick gooey substance extracted from the seeds of chaulmoogra trees by Alice Ball. It was used historically as a treatment for Hansen's disease (leprosy). It has other medical uses and is a preservative.

Decay is the condition or manner of rotting and decomposition by the activities of microbes like bacteria or fungi. It is radioactivity emitted from isotopes called decay.

Elements are atomic substances not chemically interconverted or easily broken down into simpler substances. They are the main components of all matter.

Fermentation is the biochemical conversion of nutrient substances to organic molecules, acids, bases, or gases by microorganisms like bacteria and yeasts.

Fungi are spore-making microbes that feed on organics. Fungi include molds, yeasts, and mushrooms.

Germs are microbes or microorganisms. Many germs cause diseases and are transmitted to others by fingers, inanimate objects, or contaminated foods and water.

Gerontology is the study of how people and other animals grow old.

Immune System considers how the body protects itself from dangerous germs and cancer. Two main parts are innate immunity and adaptive immunity. The immune system consists of the body's tissues, cells, molecules, and physiological pathways.

Immunology is the study of our body's protective immune system and what makes people susceptible or resistant (immune) to disease.

Inflammation is an accumulation of fluids, plasma proteins, white blood cells, and molecules caused by infection, physical or chemical injuries, or a localized immune response.

Leper Colonies were places where groups of people with Hansen's disease lived together in a quarantine-like manner.

Microbes are germs or microorganisms, like bacteria, fungi, algae, protozoa, parasites, and viruses.

Microorganisms are microbes or microscopic organisms, including bacteria, viruses, algae, protozoa, and fungi.

Nobel Prizes are bestowed to individuals for their pioneering research advances in physics, chemistry, physiology or medicine, literature, economics, and the advancement of peace. The award was established in 1900 after dynamite inventor Alfred Nobel's death.

Petri Dishes were invented by Richard Julius Petri and are shallow, circular, transparent plates with flat covers used for growing microorganisms with nutrients and agar.

Quarks are very tiny electrically charged pieces inside the non-charged neutrons of atoms.

Radioactivity is the release of ionizing radiation or particles caused by the spontaneous decay (disintegration) of atomic nuclei.

Rose Diagrams show the circular portrayal of data, first depicted by Florence Nightingale to indicate differences between clean and unclean military medical units. Nightingale used her rose diagrams to convince many to adopt hospital sanitary and cleanliness practices.

Sanitary is a clean, healthy, hygienic condition with little or no disease-causing germs.

Specimens are samples obtained from patients and used in medical testing for disease diagnosis.

Vaccines are preparations containing live, attenuated, or inactivated microorganisms, parts of microorganisms, or their disabled toxins and are meant to provoke protective immune resistance. Vaccines are used to prevent diseases.

Viruses are pathogens harboring nucleic acid genomes (DNA or RNA) enclosed inside protein coats called capsids. Viruses can multiply only while in a living cell because they do not possess the metabolic machinery for independent living.

Vital Signs are medical measurements taken from patients. They include pulse, body temperature, breathing (respiration), and blood pressure. The status of a patient's vital body functions is indicated.

Vitamin C is ascorbic acid in high amounts of citrus fruits and vegetables. It maintains healthy body cells and tissues and is a protective antioxidant.

X-rays are internal images of a patient's body. The pictures are made by passing X-rays through the body and are absorbed in varying amounts by different substances.

About the Authors

Manuel F. Varela, Ph.D., is a Professor of Biology at Eastern New Mexico University. He has a Ph.D. in biomedical sciences with an emphasis in molecular biology and biochemistry from the University of New Mexico Health Sciences Center and was a postdoctoral fellow at Harvard Medical School. His research interests include studies of bacterial multidrug efflux pumps. He has published seven books.

Ann F. Varela, M.A., is an Instructor of Mathematical Sciences at Eastern New Mexico University in Portales, New Mexico. She earned a B.S. degree and an M.A. degree from the University of New Mexico, Albuquerque, New Mexico. She has been teaching undergraduate-level mathematics and statistics courses for more than 25 years and is a member of the Phi Kappa Phi Honor Society. She is the co-author of several books, among them, *The Life and Times of the World's Most Famous Mathematicians.*

Michael F. Shaughnessy, Ph.D., is a Professor of Educational Studies at Eastern New Mexico University in Portales, New Mexico. He holds a bachelor's degree from Mercy College in Dobbs Ferry, New York, and two master's degrees and a doctorate from the University of Nebraska in Lincoln, Nebraska. He has edited, co-edited and written approximately 30 books and more than 500 articles, research pieces, and book reviews.